Logic & Mathematics
Diary of a student

Foreword

One day Stu ventured into an old bookshop located in a forgotten corner of his town. He was looking for something to enrich his knowledge and feed his insatiable curiosity. Wandering among the dusty shelves, Stu's fingers brushed against the spine of a particularly used book. Something in that little volume immediately caught his attention: it was a diary. The first page contained an introduction by Prof. Lu, who promised to explore the wonders of mathematical logic through its pages. Prof. Lu's class was a diverse group of bright young minds, each with a unique passion and a deep desire for discovery. Located in a prestigious institute dedicated to science and literature, this class was known to be a melting pot of talent and ambition.

This class represented a microcosm of how mathematics could be seen in infinite lights, each contributing their own unique lens. Together, Lu's students explored, discussed and learned, driven by their shared passion for mathematics and the inspiring guidance of their friend and classmate. Words like "dialectics", "paradoxes" and "demonstrations" danced before his eyes as Lu explained how logic was the science of reasoning and how he would use the language of mathematics to navigate through the secrets of human thought.

Prof. Lu

Hi guys, my name is Lu and together we are going to explore mathematical logic.

Mathematical logic studies the rules of thinking and reasoning using the language of mathematics.

Logic is the science of reasoning. We will study how human beings reason and use the scientific method, particularly the mathematical method, to do so. Mathematical logic focuses on mathematical reasoning, which is considered one of the most perfect forms of reasoning.

There are three main ways to study logic: dialectics, paradoxes and demonstrations. Dialectics concerns the art of discourse, paradoxes are reasoning that seems correct but leads to strange conclusions, and demonstrations are mathematical proofs that prove whether something is true or false.

We will also discuss the great thinkers of logic, such as Plato, Aristotle and Chrysippus, and how they contributed to the discipline. Mathematical logic is interesting for philosophers, mathematicians and computer scientists because it combines philosophy, mathematics and computer science.

I hope I have convinced you that mathematical logic is a fascinating subject and worth studying.

Day 01 09:00-11:00

Today, Prof. Lu started by talking about philosophy and then we will move on to mathematics and mathematical logic, until we get to computer science. Mathematical logic is interesting because it lies at the intersection of philosophy, mathematics and computer science, making it a very versatile tool.

Today we are going to talk about the liar's paradox, a very famous problem Epimenides, a Greek from the 6th century BC, said: "Cretans are liars". If Epimenides, who was Cretan, was telling the truth, then he was lying, creating a paradox. This paradox shows us that truth can be complicated.

Another example is the paradox of Eubulides, who said: 'I am lying'. If he was telling the truth, then he was lying, and if he was lying, then he was telling the truth. This paradox makes us realise that some sentences can be neither true nor false.

Day 02 15:00-17:00

Today we heard a very interesting topic: Zeno's paradoxes. We will focus on another famous paradox: that of Achilles and the tortoise.

Zeno was a Greek philosopher from the 5th century BC who tried to prove that movement is an illusion. To do this, he invented paradoxes, i.e. stories that seem to have an absurd conclusion. One of his most famous paradoxes is that of Achilles and the Tortoise.

Imagine a race between Achilles, the fastest runner, and a tortoise, the slowest animal. To make the race fairer, the tortoise starts with a 10-metre lead. Zeno says that Achilles can never overtake the tortoise because every time Achilles reaches the point from which the tortoise started, the tortoise will have moved a little further ahead. And so on ad infinitum: Achilles gets closer and closer, but never reaches it.

This paradox makes us reflect on how we think about movement and space. Zeno wanted to prove that movement is impossible because it requires crossing an infinite number of points in a finite time. Although we know that in reality Achilles would overcome the tortoise, the paradox shows us that there are interesting problems to be solved in logic and mathematics.

Zeno's paradoxes prompted philosophers and mathematicians to think more deeply about the concept of infinity and the nature of motion. They also led to the development of new ideas in mathematics, such as the theory of infinite series.

Day 03 09:00-11:00

We have talked about dialectics and paradoxes, such as the paradox of the liar and that of Achilles and the turtle. Today, however, we will focus on one of history's great mathematicians: Pythagoras.

Pythagoras was born around 570 BC and died in 490 BC. He

was one of the founders of Greek mathematics and his name is associated with the famous Pythagorean theorem. But Pythagoras was not only a mathematician; he was also a philosopher and a teacher who attracted many students.

Pythagoras taught two types of audience: 'auditors' and 'apprentices'. The auditors were like today's lecture audiences, people interested in understanding general concepts. Apprentices, on the other hand, were serious students who wanted to become mathematicians and scholars. These apprentices were called 'mathematicians', which means 'apprentices' in Greek.

Pythagoras had a very particular view of the universe. One day, while out walking, he heard the sounds of a blacksmith's hammers and noticed that some sounds were pleasant (consonants) and some were not (dissonants). He decided to investigate and discovered that consonant sounds had simple numerical relationships between hammer weights. For example, two hammers playing the same note had a ratio of 2 to 1 in their weights.

This discovery led Pythagoras to believe that everything in the universe could be explained by numbers. This is the famous Pythagorean creed: 'Everything is number'. Pythagoras believed that mathematics was the secret language of the universe, capable of describing both science and art.

Pythagoras' theorem, which says that in a right-angled

triangle, the square of the hypotenuse is equal to the sum of the squares of the other two sides, is one of his school's most famous achievements But another important discovery was that of irrational numbers. Pythagoras discovered that the diagonal of a square cannot be expressed as a ratio of integers, which came as a great shock to him and his followers.

This discovery changed mathematics forever, introducing the idea that not everything can be explained by rational numbers. The Pythagoreans swore to keep this discovery a secret, but eventually the world learned about irrational numbers.

I hope this lesson has helped you better understand the importance of Pythagoras and his discoveries.

Day 04 15:00-17:00

Today we studied Plato, a great Greek philosopher who had an enormous influence not only on philosophy, but also on mathematics. Plato was born in 428 BC and died in 347 BC, and wrote many dialogues in which he explored various aspects of human thought.

Plato believed that mathematics was fundamental to education and thought. In his dialogues 'The Republic' and 'The Laws', Plato argued that in order to become good citizens, students should learn arithmetic and geometry. According to him, these disciplines were not only useful for science and technology, but also for developing ethical and proportionate thinking, i.e. the 'right middle'.

Another interesting aspect of Plato's philosophy is his conception of nature as geometric. In the dialogue 'Timaeus', Plato describes the world as composed of perfect geometric forms, called Platonic solids. These solids are the cube, tetrahedron, octahedron, dodecahedron and icosahedron, and Plato used them to explain the structure of the universe.

Plato also contributed to logic, introducing concepts such as relative negation and the principle of non-contradiction. He defined truth as the agreement between what is said and what is actually in the world. Furthermore, he developed the theory of ideas, according to which the perfect forms of objects exist in an ideal world and their representations in our world are only imperfect shadows of these forms.

In summary, Plato combined philosophy and mathematics in a way that has profoundly influenced Western thought. His vision of mathematics as a basis for education and understanding the world is still relevant today. I hope this lecture has helped you better understand Plato's importance in mathematical logic and philosophy.

Day 05 09:00-11:00

Today, Prof. Lu spoke to us about Aristotle. Plato's philosophy was predominantly mathematical and made substantial contributions to mathematical logic. However, when one speaks of mathematical logic or logic in general, the name that comes to mind is obviously that of Aristotle.

Aristotle is still considered today, practically 2500 years later, the greatest logician who ever lived. He was a systematiser and an innovator, bringing enormous contributions that we will try to explain and review together today. One of his most important works, the Metaphysics, and the fact that the Lyceum was the school that Plato founded.

Aristotle was born in 384 BC and died in 321 BC. He was immortalised in Raphael's 'School of Athens'. At the beginning of his career as a student, Aristotle studied at Plato's Academy for 20 years, hoping to succeed him as head of the Academy. However, upon Plato's death, Aristotle did not get this role and was forced to leave the Academy.

After a period of wandering, Aristotle found work as tutor to Alexander the Great, the future conqueror. He taught Alexander Greek culture, Platonic and Aristotelian philosophy, profoundly influencing his pupil. Alexander the Great realised his master's dream, expanding Greek culture throughout the known world.

Returning to Athens in 335 BC, Aristotle founded the Lyceum, a school that became the first university and the first faculty of science. He taught practically all subjects: physics, biology, philosophy, and so on. Many of his books consist of his students' lecture notes and research papers.

Aristotle's works are enormous, almost half a million lines of work. However, only the esoteric works remain, i.e. the lecture notes and research works, while the popular works have been lost. This is unfortunate because the esoteric works are more difficult to read and understand.

One of his first major books is the Metaphysics, which contains essential contributions to logic. In particular, in Book IV of the Metaphysics, Aristotle introduces two fundamental principles: the principle of non-contradiction and the principle of the excluded third. The principle of non-contradiction states that it is not possible for the same proposition to be true and false at the same time. The principle of the third exclusion states that a proposition is either true or false, without a third possibility.

These principles form the basis of classical logic, which has become the everyday logic on which modern mathematics and science are based. Aristotle's logical works are collected under the name 'Organon', which means instrument. There are six of these works and they cover topics that are still studied in mathematical logic courses today.

The first book, 'The Categories', analyses what it means to be the subject of a proposition and what it means to be an atomic predicate. The second book, 'The Interpretation', studies compound propositions. The two most important books are 'The Analytic First' and 'The Analytic Second', which analyse arguments and the way of reasoning.

Aristotle also contributed to the study of syllogisms, quantifiers and modes. He analysed all possible types of syllogisms, discovering that there are 256, of which only a dozen and a half are correct. He also introduced rules for moving from one syllogism to another, showing that all correct syllogisms can be derived from one, the syllogism in Barbara.

Regarding quantifiers, Aristotle created a table of possible types of quantifiers: universal (all and none) and particular (some and not all). He showed that universal and particular quantifiers are interchangeable using negation.

Finally, Aristotle studied modalities (possible, impossible, necessary) and showed that they are analogous to quantifiers. For example, 'it is necessary to do something' means 'it is not possible not to do it'.

In conclusion, Aristotle is considered the greatest logician who ever lived, and his contributions to logic, such as quantifiers and modalities, are still fundamental today. However, his analysis did not reach the lowest possible level, leaving room for further developments in propositional logic and logical connectives.

Day 06 15:00-17:00

Today, Prof. Lu told us about the schools of Athens. There were three schools in Athens: Plato's Academy, Aristotle's Lyceum and a third school that we will talk about today.

The Academy was founded by Plato and among its most distinguished students was Aristotle Aristotle, after studying at the Academy for twenty years, founded his own school called the Lyceum. The third school, less well known but equally important, is the Stoà, founded by Zeno of Cyprus around 300 BC. This school took its name from the portico (in Greek 'stoà') under which the lessons were held, surrounded by paintings.

The three schools of Athens not only trained great thinkers, but also contributed to the advancement of Greek intellectual thought. An example of their importance is the diplomatic mission of 156 BC, when Athens sent three ambassadors to Rome, one for each of the three schools: the Academy, the Lyceum and the Stoa. This shows how prestigious these institutions were.

The Stoà, in particular, became famous thanks to Chrysippus of Soli, who lived between 280 and 210 BC Chrysippus is considered one of the most important logicians in history, on a par with Aristotle. Unfortunately, many of his works have been lost, but we know of his importance thanks to secondary sources such as Sextus Empiricus, who lived about 400 years after Chrysippus.

Chrysippus and the Stoics developed a different conception of logic from that of Aristotle. Whereas Aristotle saw logic as a propaedeutic tool for the sciences, Chrysippus regarded it as an autonomous science. The Stoics were the first to use variables and logical connectives such as 'not', 'and', 'or' and 'if... then', which form the basis of modern propositional logic.

An example of logical reasoning developed by the Stoics is the 'modus ponens', which states that if we have a hypothesis A and we know that A implies B, then we can conclude B.

Another example is the 'reduction to absurdity', used to show that if the contrary hypothesis leads to a contradiction, then the original hypothesis must be true.

The Stoics were also the first to define the 'functional truths' of logical connectives, describing how the truth or falsity of a proposition depends on the truth or falsity of its components. For example, a conjunction is only true if both propositions are true, whereas a disjunction is true if at least one of the propositions is true.

Finally, the Stoics anticipated the 'completeness theorem', only proved in the 20th century, which states that the syntactic rules and axioms of propositional logic are sufficient to derive all semantic truths.

In conclusion, the schools of Athens, and in particular the Stoa, have had a lasting impact on Western logic and thought. In the next lecture, we will explore the interregnum between Greek and modern logic.

Day 07 09:00-11:00

We met three great Greek philosophers: Plato, Aristotle and Chrysippus, who laid the foundations of logic.

Plato began the study of logic by opposing the Sophists and introducing the principle of non-contradiction, which states that one cannot affirm and deny the same proposition at the same time. Aristotle, considered to be one of the greatest logicians of all time, developed the theory of syllogisms, a type of reasoning that starts from major and minor premises to arrive at a conclusion. Chrysippus, on the other hand, contributed to propositional logic, which analyses propositions and logical connectives such as 'and', 'or', 'not' and 'if... then'.

After the Greek period, there were about two thousand years of development of logic, with ups and downs During the Middle Ages, logic was the focus of scholastic philosophers, who sought to bring theology closer to philosophy and science. Scholasticism, represented by figures such as William of Ockham, sought to prove faith through reason, using logical and scientific reasoning.

William of Ockham is known for his 'razor', a principle stating that entities should not be multiplied without necessity. This principle influenced modern philosophy and logic, promoting an essential and direct approach to reasoning. Ockham also studied the properties of terms, attempting to go beyond Greek logic and analysing the structure of atomic propositions.

After Scholasticism, logic continued to evolve, with figures such as Ramon Llull and Giordano Bruno who saw logic as a universal science. Llull sought to convert infidels with logic, while Bruno developed the art of memory, a technique for remembering information by arranging objects in a room.

The true precursor of modern logic was Gottfried Wilhelm Leibniz, a universal thinker who contributed in many fields, from mathematics to philosophy. Leibniz developed the idea of a universal philosophical language and rational calculus, anticipating concepts that would only be realised centuries later. His distinction between truth of reason and truth of fact has profoundly influenced modern logic and mathematics.

In conclusion, the interregnum of logic between the Greeks and modern times was a period of great development and transformation.

Day 08 15:00-17:00

Today I learnt something incredibly fascinating: how George Boole transformed the way we think about logic by introducing the idea of using mathematics to explore it! It was like being given a treasure map that reveals how ideas connect through numbers and operations, a bit like in a video game, but instead of fighting monsters, you fight with truths and falsehoods.

The journey begins in ancient Greece, where philosophers such as Aristotle and Chrysippus laid the foundation for how ideas can be logically connected. This journey through time showed me how, despite centuries of thought and debate, there was still room for revolution.

Then came the 1800s, and with it George Boole, who had this brilliant insight: what if we could use numbers to represent truth and falsehood? With "1" for true and "0" for false, he opened up a new world where logic becomes a numbers game, a bit like doing magic with mathematics.

Boole's magic did not stop at logic; it found its way into probability, theology and, most surprisingly, the technology we use every day. To think that the logic behind our computers, smartphones and even neural networks is based on Boole's ideas makes me feel like we are all a bit of magicians, using this knowledge to create things previously unimaginable.

Nevertheless, I realised that Boolean algebra has its limitations. It cannot capture every nuance of human thought or describe every kind of logic. It is a bit like when you play a video game and reach the boundaries of the game world. But that doesn't make the journey any less exciting; if anything, it makes me want to find out what lies beyond.

This journey into the world of logic and mathematics has left me with a feeling of anticipation. What other wonders are there to discover? What other heroes of logic will we meet? Today, thanks to Boole, I feel I have a compass to guide me through the mysteries of the mind and reality.

I can't wait to see where this journey will take me. But for now, I close my journal, grateful for the adventures of the mind I explored today.

Day 09 09:00-11:00

Today we explore a key figure in the history of modern logic: Gottlob Frege. Frege was born in 1848, a year of great revolutions in Europe, and lived until 1925. He is considered one of the greatest logicians of all time and made fundamental contributions to logic and mathematics.

Frege wrote three major works: 'Ideography' (1879), 'The Foundations of Arithmetic' (1884) and 'The Principles of

Arithmetic' (1893-1903). These books revolutionised the way we think about logic and mathematics. In 'Ideography', Frege created a formal language to express pure thought, realising Leibniz's dream of a universal language for the sciences. This formal language was the first step towards modern logic.

Frege introduced two fundamental concepts that made a big step forward compared to the logic of the Greeks and Boole: relations and quantifiers. Aristotle's logic was based on subject and predicate, but Frege extended this analysis to include more complex relations, such as equality, major and minor, which are essential in mathematics. In addition, he introduced quantifiers, which allow statements such as 'for every number there is another greater'.

In his second book, The Foundations of Arithmetic, Frege attempted to reduce arithmetic to logic. He showed that numbers can be defined in logical terms, for example, zero as the empty set and one as the set containing only the empty set. This attempt to reduce arithmetic to logic was an important step towards the logical foundation of all mathematics.

However, in 1902, Bertrand Russell discovered a paradox that challenged Frege's work. Russell's paradox concerns the set of all sets that do not belong to themselves and showed that this set leads to a contradiction. Frege failed to solve this paradox, but his work nevertheless laid the foundation for the development of modern logic.

Today, set theory is based on the Zermelo-Fraenkel axioms, which avoid Russell's paradox. However, Frege's dream of a complete logical foundation of mathematics was not realised, and Gödel's incompleteness theorem proved that no list of axioms can be exhaustive.

The lecture on Frege concluded by emphasising the profound impact of his work on logic and mathematics. His attempt to give a logical foundation to arithmetic, while not completely successful, paved the way for future research and continues to be a landmark in modern thought.

Understanding Frege's contributions helps us appreciate the depth and complexity of mathematical logic, a field that is as abstract as it is fundamental to the advancement of science and technology.

Day 10 15:00-17:00

Today's lesson was particularly fascinating and enlightening as we explored the life and achievements of one of the giants of mathematical logic, Bertrand Russell. Illustrating his existence is no easy task, given the breadth of his accomplishments, both professional and personal. Russell's life, which spanned almost a century (1872-1970), was as extraordinary in its intellectual contributions as in its personal experiences.

Russell, an emblematic figure known for his outstanding contributions to logic, was also a philosopher, political writer, activist and Nobel laureate in literature in 1950. His work crossed disciplinary divides, profoundly influencing philosophy, literature, politics and of course, logic and mathematics. Russell's paradox remains a milestone in the field of logic, illuminating the foundations of set theory and challenging mathematicians to reconsider some of their basic assumptions.

Beyond the academic sphere, Russell was notable for his progressive views on issues of morality and society, as discussed in his influential book 'Marriage and Morality'. He was an ardent advocate of premarital cohabitation and the use of contraceptives, ideas that were particularly radical for the time.

Russell's anti-war activism saw him emerge as a central figure in the 1960s, particularly with his opposition to the Vietnam War and the founding of the Russell Tribunal. His actions catalysed public opinion and highlighted the horrors committed during the conflict.

His autobiography, divided into three volumes, reveals not only his achievements and failures, but also his relationships with other influential figures of the time, including Freud and Einstein. It is a testament to his extraordinary mind and complex personal life.

Bertrand Russell's complexity and influence persist. Today's lectures have underlined not only the magnitude of his contributions, but also the way he continues to inspire debate and reflection today. His writings, especially 'My Life in Philosophy' and 'Introduction to the Philosophy of Mathematics', remain lectio divina for anyone fascinated by the intersection of logic, philosophy and life. The lessons learned today about Russell's centrality to 20th century thought and his enduring legacy will not be forgotten. End of lecture today and thoughts to ponder until our next meeting.

Day 11 09:00-11:00

Today in class we delved into a character who, despite being less exposed in the media than his illustrious mentor Bertrand Russell, proved to be of incredible depth and charm: Ludwig Wittgenstein. He lived in a period of profound intellectual revolutions, in which he played a non-marginal role.

Wittgenstein's story is that of a man born into a family of exceptional wealth and culture in late 19th century Vienna. His family life, surrounded by leading figures in the arts such as Gustav Klimt and close ties to musical personalities such as Maurice Ravel, reveals the fertile soil from which his thought sprouted. The historical coincidence of his school days with Adolf Hitler in Linz adds drama to the context of his education.

He made several significant moves, and his move to Cambridge to learn mathematical logic directly from Bertrand Russell marked the beginning of an intellectual journey of extraordinary originality. Despite Russell's appreciation, Wittgenstein's search for truth pushed him off the beaten track until he retired to Norway to reflect in solitude.

His service and imprisonment in the First World War were times of trial but also of prolificacy, during which he matured the 'Tractatus Logico-Philosophicus'. This work, despite its brevity, reflected the depth of his thinking on the correspondence between language, thought and reality.

He then became a primary school teacher, a period during which doubts emerged about the completeness of his earlier philosophical conclusions. This turmoil led him to a return to Cambridge and a two-decade commitment that culminated in the 'Philosophical Researches', where he introduced the revolutionary concept of 'language games'.

His innovative vision of language and its use materialised an immense legacy to philosophy, despite Frege's admission of his work as more art than science. Today's lecture opened a window not only on the life of a great thinker but also on his lasting impact on the philosophy of language and beyond.

Day 12 15:00-17:00

Today, we addressed an extremely interesting argument concerning mathematical logic and the remarkable contribution made by two eminent mathematicians, Hilbert and Brouwer. Our attention has specifically turned to Hilbert, who lived between 1862 and 1943, and whose achievements have left an indelible mark on mathematics and logic.

Hilbert is famous for his axiomatic method applied to mathematics, which aims to construct the discipline from a defined set of axioms. Axioms are essentially primary rules or principles on which all further theoretical demonstrations are based; a classic example of axioms can be found in Euclidean geometry, with its points, lines and planes.

One of the greatest challenges faced by Hilbert was the search for the consistency of axioms. Such consistency is essential to prevent contradictions, ensuring that the truths proved cannot be simultaneously true and false. Hilbert employed this concept to prove the consistency of Euclidean geometry, showing that if an alternative form of geometry, such as hyperbolic geometry, was consistent, so would Euclidean geometry be.

His thinking did not stop there. Hilbert outlined a list of 23 unsolved problems in 1900, during a congress in Paris, establishing a sort of 'to-do list' for subsequent generations of mathematicians. Among these problems was the question of the consistency of mathematical analysis, on which much of modern mathematics is based.

Hilbert strove to build a coherent mathematical framework free of contradictions, drawing on clearly defined axioms from which theorems could be rigorously derived. The exceptional impact of his work on mathematics and logic persists to this day, with many of the problems he posed continuing to be studied.

Day 13 09:00-11:00

Today, Prof. Lu told us about a truly extraordinary figure in the field of mathematical logic: Kurt Gödel. Gödel is considered one of the greatest logicians of all time, on a par with Aristotle. His life and work have had an enormous impact on mathematics and philosophy.

Gödel was born in 1906 and lived until 1978. His career can be divided into two main parts: the first part in Vienna and the second part in Princeton, USA. In Vienna, Gödel was part of the famous Vienna Circle, a group of philosophers and scientists who dealt with logic and the philosophy of science.

Here, under the influence of people like Rudolf Carnap, Gödel began working on fundamental problems in mathematical logic.

One of his first major achievements was the completeness theorem, proved in 1930. This theorem states that all logical truths can be proved within a formal system, such as the one developed by Frege. However, the result that made him truly famous was the incompleteness theorem, proved in 1931. This theorem revolutionised mathematics by proving that in any sufficiently powerful formal system there are truths that can neither be proved nor disproved within that system. In other words, there are always mathematical truths that escape our ability to prove.

Gödel also proved that a formal system cannot prove its own consistency, i.e. it cannot prove that it does not contain contradictions. This is a bit like saying that a person cannot prove on his own that he is not insane. Only an inconsistent, i.e. 'crazy', system could claim to be consistent.

After leaving Vienna due to the Nazi invasion, Gödel moved to Princeton, where he worked at the Institute for Advanced Study alongside great minds such as Albert Einstein. At Princeton, he continued to do research in various fields, including set theory and general relativity. For example, he proved that the continuum hypothesis, a fundamental problem in set theory, can neither be proved nor disproved by the means of modern mathematics.

Gödel also became interested in general relativity and proved that in certain models of the universe it is possible to travel backwards in time. This result, although theoretical, has fascinated many physicists and philosophers.

In the last years of his life, Gödel devoted himself to theology, trying to reformulate the ontological proof for the existence of God in mathematical terms. Although he did not personally believe in God, he wanted to prove that it was possible to construct a rigorous logical argument for the existence of a supreme being.

Gödel's life and work show us the limits and wonders of logic and mathematics. His incompleteness theorems remind us that we can never know everything and that there will always be mysteries to explore. Gödel broke new ground in mathematics and philosophy, and his work continues to influence these fields today.

Day 14 09:00-11:00

Today we talked about a very interesting topic: Gödel's incompleteness theorem. This theorem is one of the most important results of mathematical logic and tells us something very profound about mathematics and knowledge in general.

Imagine you have a book that tells a story. In mathematics, this book is like a set of rules or axioms from which we can deduce other truths. But just as a book cannot describe every single detail of the story, a mathematical system cannot contain every possible truth. Gödel showed that in any sufficiently powerful mathematical system (like the one we use for integers), there will always be truths that can neither be proved nor disproved within that system.

To understand better, think of a sentence that says 'I am not demonstrable'. If this sentence were provable, then it would be false, which is a contradiction. If it is not provable, then it is true, but we cannot prove it. This creates a situation where we know that the sentence is true, but the mathematical system cannot prove it.

Gödel also proved that a mathematical system cannot prove its own consistency, i.e. it cannot prove that it does not contain contradictions. This means that we can never be completely sure that a mathematical system is error-free, which is a bit like saying that a person cannot prove on his own that he is not insane.

Gödel's theorem shows us the limits of mathematics and knowledge: we can never know everything and we can never be completely sure that we are not wrong. But this should not discourage us; on the contrary, it encourages us to continue exploring and searching for new truths!

Day 15 15:00-17:00

Today, Prof. Lu talks to us about truth in logic. Imagine yourself in a trial, like that of Jesus before Pontius Pilate. Pilate asked Jesus, "What is truth?" and this question has become central in philosophy and logic.

To understand better, let us start with a famous paradox: the liar's paradox. If I say 'I am lying', is this sentence true or false? If it is true, then I am lying, so it is false. But if it is false, then I am telling the truth, so it is true. This paradox shows that some sentences can be neither completely true nor completely false.

Plato and Aristotle, two great Greek philosophers, tried to define truth. Plato said that a sentence is true if what it says corresponds to reality. For example, 'snow is white' is true if, in the real world, snow is indeed white. Aristotle added that a sentence is false if what it says does not correspond to reality.

Over time, logicians have developed methods for analysing more complex sentences using logical connectives such as "and", "or", "not" and "if... then". For example, a sentence such as 'the snow is white and the sky is blue' is only true if both statements are true.

Finally, Tarski, a 20th century logician, proposed a modern solution: truth can only be defined in a metalanguage, i.e. a language that speaks of language itself. This means that we can only say whether a sentence is true or false if we use a broader language that includes all possible interpretations.

Truth is a complex, but fascinating subject!

Day 16 09:00-11:00

Today we deal with a central theme in mathematical and philosophical logic: the problem of truth. Do you remember that we started our course by talking about one of the most important paradoxes, the paradox of the liar? This paradox was fundamental to the development of logic and mathematical logic.

The liar's paradox is a sentence that says: 'I am lying'. If this sentence is true, then he is telling the truth, so it should be false. But if it is false, then he is telling the truth. This creates a contradiction and shows us that there are sentences that can be neither true nor false. This paradox challenged the foundations of classical logic, which is based on the principle of non-contradiction and the principle of the excluded third.

Now, let us try to answer Pilate's question: "What is truth?" To do so, we need to look at how philosophers and logicians have dealt with this question throughout history. Let us start with the Greek philosophers, in particular Plato and Aristotle.

Plato, in his dialogue 'Sophist', gave a definition of truth that has become very famous. According to Plato, a sentence is true if there is a correspondence between what is said in language and what happens in the world. For example, the sentence 'Snow is white' is true if, and only if, snow is actually white in the world. This is a definition of truth based on the correspondence between language and reality.

Aristotle took this definition and made it more precise. In his 'Metaphysics', he said that it is true to say of what is that it is, and of what is not that it is not. It is false to say of what is that it is not, and of what is not that it is. In other words, a sentence is true if it describes reality correctly and false if it describes it incorrectly.

Let us now turn to modern logic. One of the most important

contributions was made by Alfred Tarski in the 1930s. Tarski showed that it is not possible to define truth within a formal language, but only in a metalanguage. He introduced the distinction between language and metalanguage, where metalanguage is the language in which we speak about language. For example, when we learn English, we speak in Italian with our teacher. Italian is the metalanguage and English is the language.

Tarski gave a formal definition of truth for formal languages, such as those of mathematics and logic. He showed that a sentence such as 'Snow is white' is true if, and only if, the snow is indeed white. However, this definition only works for formal languages and not for natural languages such as Italian or English.

There have been other theories of truth proposed by philosophers such as Austin and Kripke Austin introduced the idea that truth depends on situations. A sentence can be true in one situation and false in another. Kripke, on the other hand, spoke of 'landing', i.e. the fact that sentences must be connected to reality in order to have a truth value.

In conclusion, the problem of truth is complex and has no single solution. Tarski's definition is very useful for formal languages, but for natural languages there are other theories that try to explain how truth works.

Day 17 15:00-17:00

Today we study one of the most fascinating figures in mathematical logic and computer science: Alan Turing. The lesson is entitled "The Enigma of Computer Science" because Turing had a life full of mysteries and made discoveries that revolutionised the world.

Today, we will focus on the incredible contributions to science by Alan Turing (1912-1954). Turing worked on many different topics, from the invention of the first computers to espionage and the study of living organisms.

One of his most important contributions is the 'Turing machine'. Between 1936 and 1939, Turing asked himself what was possible to calculate mechanically. He invented a theoretical model of a calculator that could perform any calculation that a human being could do by following precise rules. This model became the basis for modern computers.

During the Second World War, Turing worked for British espionage. He deciphered coded messages from the Germans, who used a machine called Enigma. Thanks to his work, the Allies could predict the moves of the Germans and this had a huge impact on the outcome of the war.

After the war, Turing contributed to the birth of computer science. He imagined that computers could do more than just calculations. In 1950, he proposed a revolutionary question: "Can machines think?" He introduced the famous 'Turing test' to determine whether a machine can be considered intelligent. If a machine can answer questions indistinguishably from a human being, then it can be considered intelligent.

Turing also explored artificial intelligence and morphogenesis, the study of how living organisms develop their forms. He sought to understand how information in DNA can create complex structures such as plants and animals.

In conclusion, Alan Turing was a pioneer in many fields and his ideas continue to influence science and technology today. His life and work show us how important it is to be curious and try to understand the secrets of the world around us.

Day 18 09:00-11:00

We have already explored many important figures in the history of logic, from the Greeks to modern times. But what has happened in recent times in mathematical logic and how has this discipline evolved into an independent branch of modern mathematics?

Mathematical logic today is divided into four main branches: model theory, proof theory, recursivity theory and set theory. Each of these branches has its origins in what we have already seen in the past, but developed further during the 20th century.

Let us begin with the theory of patterns. When we talk about 'patterns', we do not mean fashion catwalks, but the study of semantics, i.e. the meaning of mathematical signs and formulae. Pattern theory is concerned with understanding what formulas mean and how we can interpret them. This branch originated with the work of Alfred Tarski, who introduced a formal definition of truth in 1936. Tarski showed that we cannot define truth within a formal system, but only in a metalanguage, i.e. a language outside the system itself.

Another important figure in model theory is Abraham Robinson, who in the 1950s and 1960s developed non-standard analysis, a version of mathematical analysis that uses infinitesimals, concepts that had been set aside for centuries. Robinson also contributed to the study of algebraic structures through logic.

Let us now turn to demonstration theory, which deals with syntax, i.e. signs and written formulae. This branch originated with Gerhard Gentzen, who in 1936 demonstrated the consistency of arithmetic using a transfinite induction principle. Demonstration theory then evolved through the work of logicians such as Kurt Gödel, who proved that we cannot prove the consistency of a formal system within the system itself.

In the 1950s and 1960s, logicians such as Kurt Schütte and Takeuti tried to prove the consistency of analysis, the theory of real numbers. In the 1980s and 1990s, Jean-Yves Girard proved the consistency of analysis in a non-constructive way, paving the way for new research in proof theory.

The third branch is recursivity theory, which studies the potential and limitations of computers. This theory originated with Alan Turing, who in 1936 introduced the Turing machine, an abstract model of a computer. Recursivity theory is concerned with understanding what is possible to calculate and what is not. People such as Stephen Kleene and Gerald Sacks have contributed to the development of this theory, studying calculability on integers and real numbers.

Finally, we have set theory, which studies collections of objects. This theory was developed by Georg Cantor in the late 19th century and led to the discovery that there are different types of infinity. Cantor proved that the real numbers form an infinity larger than that of the integers. The continuum hypothesis, one of the most famous problems in set theory, concerns the question of whether an intermediate infinity exists between that of the integers and that of the real numbers.

Kurt Gödel and Paul Cohen proved that the continuum hypothesis is independent of set theory, i.e. it can neither be proved nor disproved within this theory. This result had a great impact on mathematical logic and led to new research into the axioms of infinity and the determination of infinite games.

I hope this explanation has helped you better understand the different branches of mathematical logic and their recent developments Next time, we will talk about the influences of logic on the foundation of mathematics.

Day 19 15:00-17:00

Today we have explored a crucial topic: the fundamental importance of mathematical logic as the basis of all mathematics. Throughout the centuries, many illustrious mathematicians have pursued the goal of basing mathematics on firm and irrefutable principles, a journey that begins with our Greek ancestors and runs throughout history up to the 20th century.

We began our time journey with Pythagoras, in the 6th century BC, and his belief that every element could be represented by integers, until he stumbled upon the paradox of irrational geometric quantities, triggering one of the first great crises in mathematics.

Later, in the 3rd century BC, it was Euclid who gave a new direction to mathematics, founding it on geometry in his work 'The Elements'. Euclid's perspective remained unchallenged for over two thousand years.

In the 17th century, Descartes revolutionised the conception of geometry with the introduction of Cartesian geometry, linking geometry and real numbers in an unprecedented way.

The 19th century saw Dedekind address the issue of irrationals through the theory of infinity, allowing arithmetic and geometry to be reconciled.

The 20th century was marked by the foundations crisis caused by Russell's paradox, which stimulated new foundations for mathematics, focusing on set theory, structures, categories and lambda calculus.

We conclude by acknowledging that, despite the success of set theory as a fundamental approach to mathematics, any foundation will inevitably show incompleteness. Mathematical logic, however, continues to provide us with a solid ground for mathematical reasoning. This final lecture has hopefully been a key to a better understanding of the immense value of mathematical logic in the vast and exciting universe of mathematics.

Epilogue

As Stu closes the diary, he thinks of the wide panorama he has traversed with those pages, from the arid lands of ancient Greece to the endless fields of modernity. Travelling through epochs and minds, he observes how three giants - Plato, Aristotle and Chrysippus - laid a solid foundation on which logic rose, revealing its power and beauty.

Plato confronted the Sophists with his principle of non-contradiction, establishing a solid foundation for truth. Aristotle, then, expanded the domain of logic with his syllogisms, allowing us to navigate through the complexity of premises to the bright shores of conclusions. Chrysippus, with his propositional logic, wove a web that captured the various forms in which truth and falsity manifest themselves in our propositions.

The tenacity of medieval scholars, such as William of Ockham, who, armed with the razor of simplicity, cut through the complexities in pursuit of pure, rational faith. They demonstrated that logic is not just an intellectual exercise, but a tool for understanding the divine.

Throughout the flow of time, logic has continued to evolve, crossing the boundaries of philosophy to become the universal language of mathematics and science. Leibniz's vision of a universal philosophical language and rational calculation laid the foundation for modern computing, transforming the way we live, work and think.

And now, at the end of our journey, we find ourselves in a world where logic permeates every aspect of our existence, from the devices we use every day to the theories that define our understanding of the universe. Logic, from a simple tool of argumentation, has become a mainstay of modern civilisation.

Logic will continue to evolve, to challenge and be challenged, as we, its eternal students, try to grasp its mysteries.

www.ingramcontent.com/pod-product-compliance
Lightning Source LLC
Chambersburg PA
CBHW072056230526
45479CB00010B/1105